Nature School

LET'S LOOK FOR BEETLES!

By Seth Lynch

Gareth Stevens Publishing

Please visit our website, www.garethstevens.com. For a free color catalog of all our high-quality books, call toll free 1-800-542-2595 or fax 1-877-542-2596.

Library of Congress Cataloging-in-Publication Data

Names: Lynch, Seth, author.
Title: Let's look for beetles! / Seth Lynch.
Description: Buffalo, New York : Gareth Stevens Publishing, [2024] |
Series: Nature school | Includes index. | Audience: Grades K-1
Identifiers: LCCN 2022045110 (print) | LCCN 2022045111 (ebook) | ISBN 9781538286173 (library binding) | ISBN 9781538286166 (paperback) | ISBN 9781538286180 (ebook)
Subjects: LCSH: Beetles–Juvenile literature.
Classification: LCC QL576.2 .L96 2024 (print) | LCC QL576.2 (ebook) | DDC 595.76–dc23/eng/20220920
LC record available at https://lccn.loc.gov/2022045110
LC ebook record available at https://lccn.loc.gov/2022045111

First Edition

Published in 2024 by
Gareth Stevens Publishing
2544 Clinton St.
Buffalo, NY 14224

Editor: Kristen Nelson
Designer: Andrea Davison-Bartolotta

Photo credits: Cover Yenyu Shih/Shutterstock.com; p. 1 forest71/Shutterstock.com; p. 5 (top left, bottom left) Mark Brandon/Shutterstock.com; p. 5 (top right) MLReed/Shutterstock.com; p. 5 (bottom right) Noheaphotos/Shutterstock.com; p. 7 Darkdiamond67/Shutterstock.com; p. 9 Vinicius R. Souza/Shutterstock.com; pp. 11, 24 (egg) Kuttelvaserova Stuchelova/Shutterstock.com; pp. 13, 24 (leg) MakroBetz/Shutterstock.com; pp. 15, 24 (wing) I Wayan Sumatika/Shutterstock.com; p. 17 Suede Chen/Shutterstock.com; p. 19 Suzanne Tucker/Shutterstock.com; p. 21 Japan's Fireworks/Shutterstock.com; p. 23 Per-Boge/Shutterstock.com.

Printed in the United States of America

CPSIA compliance information: Batch #CS24GS: For further information contact Gareth Stevens, New York, New York at 1-800-542-2595.

Contents

Let's look for beetles! There are many sizes and colors.

Ladybugs are red
or yellow.
They have spots.

June bugs are brown.
They like warm weather.

Beetles are insects.
They lay eggs.

They have six legs.

They have wings.
Most beetles fly.

Ground beetles
do not fly.

It is getting dark.
I see fireflies!

They are beetles. Their light can be yellow, green, or orange.

What other beetles can we find?

Words to Know

egg

leg

wing

Index